EASY ANSWERS

TO FIRST SCIENCE QUESTIONS ABOUT

SPACE

WRITTEN BY Q. L. PEARCE

ILLUSTRATED BY GIL HUNG

EXPERT CONSULTANT: Jonathon Hodge, Planetarium Director,
Santa Monica College, Santa Monica, California

TO LAURA AND LYNN

An RGA Book

Copyright © 1991 by RGA Publishing Group, Inc.
This edition published in 1991 by SMITHMARK Publishers Inc.,
112 Madison Avenue, New York, NY 10016.
Manufactured in the United States of America.

ISBN 0-8317-2587-7

Q: *How big is the universe?*

Answer: The universe is larger than anything you can imagine. In fact, no one knows how big it is. It includes everything that we know about and things we haven't discovered yet. On a clear, dark night, in a place far from city lights, you can see about 4,500 stars. But these are only a few of the *billions* of stars that make up the Milky Way, the galaxy that is our home. And the Milky Way is just one of *billions* of galaxies that fill our vast universe.

Q: WHY IS SPACE BLACK?

Answer: Space looks black because it is almost empty. Earth's air is made up of very tiny particles, or bits, called molecules. When light from the Sun passes through the air, it strikes these particles and is scattered all around. Sunlight is made up of different colors of light. The blue and violet colors scatter the most, so we usually see the sky as blue. Because there are no air molecules in space to scatter sunlight, we see space as black.

Q: *Is it cold in space?*

Answer: Space doesn't have any temperature at all. Temperature is a measurement of the movement of particles, and space does not contain enough particles for it to have a measurable temperature. However, an *object* in space is made up of particles, so it can be hot or cold. The actual temperature of an object in space depends on how far it is from, or if it faces, the Sun. For example, the side of a spacecraft facing the Sun is warm, and the side facing away from it is cold.

Q: CAN YOU HEAR SOUND IN SPACE?

Answer: No. Sound is the result of the movement, or vibration, of molecules. When something makes noise, such as a clanging bell, its molecules move. These moving molecules bump air molecules, causing them to move, too. The vibrations (or sound waves) travel through the air and strike the part of your ear called the eardrum. You then "hear" the vibrations as sound. But there is no air in space to carry sound waves to your ear, so you cannot hear sound in space.

Q: WHY ARE THINGS IN SPACE WEIGHTLESS?

Answer: Things in space are weightless because they are falling and there is nothing to resist their fall. Gravity is the invisible force of an object that pulls, or attracts, every other object to its center. You feel the force of gravity as weight when something (such as your hand) resists the fall of an object (such as a brick) toward the center of the Earth. When you jump from a diving board, you are weightless until the water stops your fall. Astronauts in an Earth orbit feel weightless because their ship is "falling" in a curved path around the planet.

Q: HOW ARE STARS FORMED?

Answer: Stars form from huge, spinning clouds of gas and dust. Most of the gas is hydrogen. As a cloud spins, it begins to shrink down into a flattened disk that looks something like a Frisbee with a thick bump in the center. The center becomes very hot because the gas there is being squeezed into a dense ball. When the ball is hot enough, the star ignites, or "turns on," and begins to burn its gas fuel.

Q: WHY DO STARS TWINKLE?

Answer: Stars seem to twinkle because you see them through a thick layer of air. Earth's air moves and ripples; it is rarely still. As starlight passes through warm and cold layers of unsteady air, it bends and so appears to quiver or twinkle. On a very windy night, even planets, which are much closer than the stars, can appear to twinkle.

Q: WHERE DO STARS GO DURING THE DAY?

Answer: The stars you see at night are still there during the day. You just can't see them in the daytime because the light from the Sun is brighter than the starlight. In the same way, it is more difficult to see the glow of a flashlight in the daytime than it is at night. When the Sun sets in the evening, the sky darkens and the stars slowly appear. When the Sun rises in the morning, the stars fade from sight.

Answer: Yes. The Sun is an average-sized star. Many stars are bigger and brighter than the Sun. The reason our Sun appears to be the biggest and brightest star in the sky is that we are much closer to it than we are to other stars. Most of the stars we see at night are at least a *million* times farther away than the Sun! Our yellow Sun is also average in temperature. The coolest stars are red and the hottest stars are white or blue.

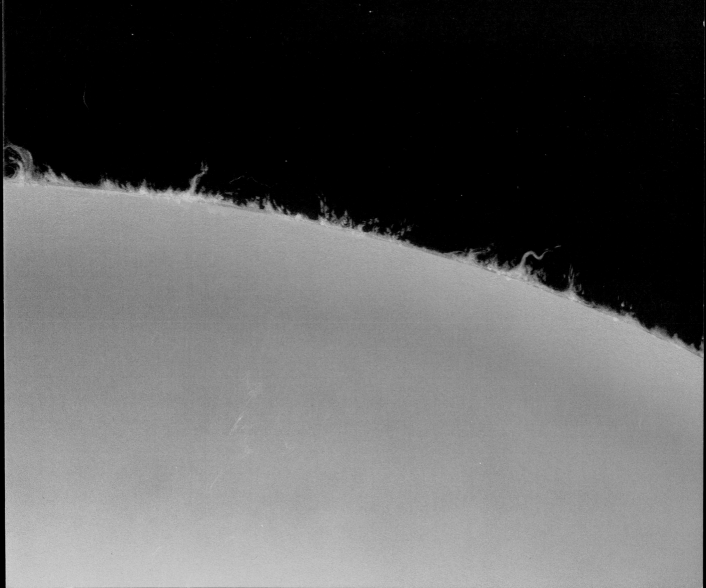

Q: WILL THE SUN SHINE FOREVER?

Answer: No. Stars like ours burn for about 10 billion years. Our Sun is now half that age. From its birth, the Sun has shone by breaking down hydrogen atoms and reassembling them into helium atoms. When it uses up its hydrogen, the Sun will no longer shine. It will first swell and become a red giant star, then collapse and become a white dwarf star. It will then fade to a cold, black body in space. But this won't happen for at least 5 billion years!

Answer: As the Earth circles the Sun, our view of star groups, called constellations, changes. To demonstrate this, walk in a big circle. After every few steps, look directly over your right shoulder. Your view will keep changing until you are back where you started. Every 365¼ days, the Earth completes one trip around the Sun. We see different parts of the starry night sky at different times in that journey.

Q: *IF THE EARTH IS MOVING, WHY CAN'T WE FEEL IT?*

Answer: You don't feel the Earth moving because you and everything around you are moving *with* it. The Earth moves in space in several different ways. It spins on its axis (an imaginary line from the North Pole to the South Pole) at a rate of about 1,000 miles per hour (mph). It also orbits around the Sun at nearly 70,000 mph. Last, the Sun wheels around the galaxy at about 600,000 mph, carrying Earth and the other planets with it.

Q: WHERE DO COMETS COME FROM?

Answer: Comets come from far out in deep space. They are small, icy chunks of frozen gas and dust. Astronomers believe that a huge cloud of billions and billions of comets, called the Oort cloud, surrounds our Solar System. Sometimes a comet from this cloud falls into a closer orbit around the Sun. As it draws near the Sun, it warms up. Some of the ice boils away, forming a halo around the comet (called a coma) and a glowing tail that can be seen in the night sky.

Q: WHAT ARE ASTEROIDS MADE OF?

Answer: Asteroids are made of rock and metal probably left over from the birth of the Solar System. Astronomers have charted several thousand asteroids. The largest, Ceres, is nearly as wide as the state of Texas. Most are much smaller. If all the known asteroids were gathered together into one planet, that planet would still be smaller than our Moon. Most asteroids tumble around the Sun in orbits between Mars and Jupiter.

Q: WHAT IS A MOON?

Answer: A moon, also called a satellite, is a body in space that orbits around a planet. A moon does not shine by its own light. Instead, it shines when sunlight is reflected, or bounced, off its surface. Some of the satellites in our Solar System took shape as their planets developed. They formed in the places where we see them today. Other moons were pulled by a planet's gravity into orbit. It is likely that the two moons of Mars are asteroids captured by gravity when they came too close to the planet.

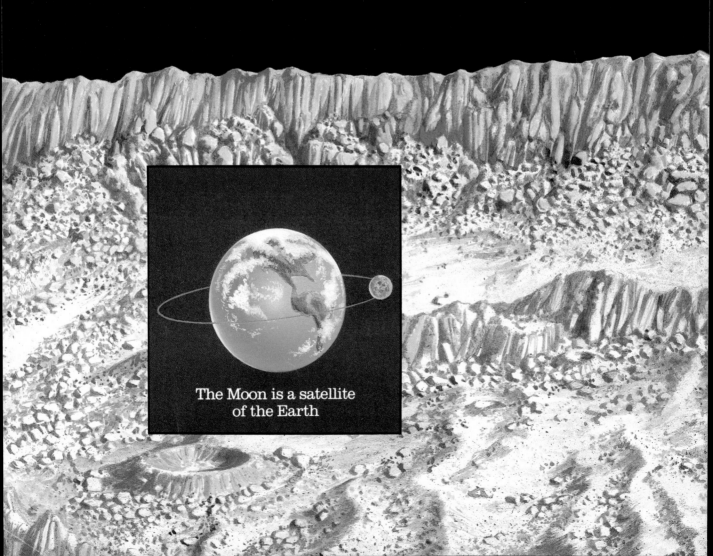

The Moon is a satellite
of the Earth

Q: WHY DOES THE SAME SIDE OF OUR MOON ALWAYS FACE THE EARTH?

Answer: We always see the same side of the Moon because of the way it moves. The Moon rotates once on its axis every $27\frac{1}{3}$ days. That is the same time that it takes to complete one orbit around Earth. To test this, pretend that you are the Moon. Using a chair as the Earth, stand with your left shoulder toward it. Walk around the chair, keeping your shoulder toward the chair, until you are back where you started. You have completed one orbit and one rotation, facing each wall in the room in turn, while your left shoulder always faced "Earth."

Q: Do all planets have moons?

Answer: No. Two of the nine planets in our Solar System, Venus and Mercury, do not have moons. Earth and distant Pluto have one moon each, and two moons circle Mars. The giant planets have many moons. Neptune has eight known moons, Uranus has fifteen, and Jupiter has sixteen. Saturn has the most moons with at least seventeen. The giant planets probably have more small moons that we haven't discovered yet!

Q: WHICH IS THE LARGEST MOON IN THE SOLAR SYSTEM?

Answer: The lunar giant of the Solar System
is Ganymede of Jupiter. Some 3,280 miles wide,
Ganymede is nearly matched in size by Saturn's
moon Titan, which is about 3,200 miles wide. Both
moons are larger than the planets Mercury and Pluto.
Grooves, ridges, and craters cover the icy surface of
Ganymede. The surface of Titan is a mystery, because it
is blanketed by orange-tinted clouds. Titan, in fact, is the
only moon in the solar system that has an atmosphere

Q: *DO OTHER PLANETS HAVE OCEANS?*

Answer: Earth is the only planet in the Solar System with huge oceans of water. Scientists think that Venus may have had oceans millions of years ago. If so, the Venusian oceans disappeared as the planet heated up to scorching temperatures. On Mars, rivers of water may have once carved some of the channels we see, then flowed into small Martian seas. If there is any water on Mars now, it lies frozen beneath the planet's dry surface.

Q: ARE THERE VOLCANOES ON OTHER PLANETS?

Answer: Earth probes that visited Venus found evidence of several volcanoes and lava-covered plains on that planet. The largest volcano in the Solar System is the Martian volcano Olympus Mons. Very old and inactive, Olympus Mons is three times larger than the biggest volcano on Earth. Volcanoes rise not only from the surfaces of planets, but from the surfaces of moons, too. Io (EYE-oh), one of Jupiter's moons, has at least ten active volcanoes.

Q: *WHICH PLANET IS FARTHEST FROM THE SUN?*

Answer: Pluto is usually the most distant planet in the Solar System, but sometimes Neptune is. Pluto's average distance from the Sun is more than 3 billion miles. This tiny planet has an unusual orbit that sometimes crosses inside the path of the planet Neptune. Because of this, since 1979 Neptune has been farther from the Sun than Pluto. By 1999, however, the two orbits will cross again, and Pluto will regain its place as the most distant planet.

Q: *WHICH IS THE LARGEST PLANET?*

Answer: Jupiter, nearly eleven times wider than the Earth, is the largest planet in our Solar System. It is about 89,000 miles across. That is more than one-third of the distance from the Earth to the Moon! Because of its size, you can easily see Jupiter in the night sky. This colossal planet is five times farther from the Sun than the Earth is. When light leaves the Sun, it takes about eight minutes to reach Earth, and more than forty minutes to reach Jupiter.

Jupiter

Q: WHY IS NEPTUNE BLUE?

Answer: Neptune appears blue because of a gas in its atmosphere called methane. Methane absorbs red light and reflects blue, giving Neptune its soft bluish hue. Neptune is the outermost of the giant gas planets. The others are Jupiter, Saturn, and Uranus. Neptune is about 3 billion miles from the Sun, and its year is 164 Earth years long. In fact, Neptune hasn't even completed one full orbit since its discovery in 1846.

Q: WHY IS URANUS TIPPED ON ITS SIDE?

Answer: Scientists aren't yet certain why Uranus, with all its moons and narrow rings, is tipped on its side. Perhaps a huge object struck Uranus soon after it formed and "knocked" the planet on its side. If so, it must have been a very violent collision! Whatever the reason, Uranus is certainly different from the other planets. Its orbit is a full eighty-four years. For half of that time, its South Pole directly faces the Sun. The other half of the time Uranus's North Pole faces the Sun.

Answer: Saturn's rings are made of ice and rock. Some of the rocks are as big as trucks, while others are merely grains of dust. Most are about four to six feet wide. Saturn is not the only planet that has rings, but its rings are the most spectacular. The rings are actually composed of hundreds of tiny ringlets that extend more than 40,000 miles out into space. The rings may be all that remains of a moon that was destroyed in a great collision.

Q: WHY DOES MARS APPEAR RED?

Answer: Mars appears red because rust-colored rocks and dust cover its surface. Have you ever seen an old bicycle that has been left out in the rain? The metal parts become covered with rough, red-orange rust. This rust is also called iron oxide. The sand and dust of Mars contain plenty of iron oxide. The Martian wind lifts the sand and dust into the air, so that even the sky looks reddish-pink.

Q: *Is Mercury always hot?*

Answer: No. The surface of Mercury that is turned away from the Sun is bitterly cold. Although Mercury is the closest planet to the Sun, it has little atmosphere to hold in heat. Mercury's dry desert surface reaches temperatures of up to 800 degrees Fahrenheit during the day. That's about twice as hot as an oven! At night, however, the planet's surface loses its heat to space, and the temperature drops to a shivery −345 degrees. That's much colder than a freezer!

Q: WHY IS VENUS CALLED OUR "SISTER PLANET"?

Answer: Earth and Venus are called sister planets because they are similar in size and weight. Both receive about the same amount of energy from the Sun, and they probably formed in the same way. Venus and Earth may be sisters, but they are not twins. The atmosphere of Venus is made up mostly of a gas called carbon dioxide. This gas holds in heat, making Venus up to a scorching 900 degrees. Also, the air on Venus is ninety-five times thicker than the air on Earth.

Earth

Venus

Q: Is Earth the only planet with life?

Answer: Earth is the only planet in our Solar System that supports life as we know it. That's because Earth is the right temperature and distance from the Sun for life to exist here comfortably. Earth's atmosphere is not only breathable, but it also shields the planet's surface from the Sun's harmful rays. Living things need water to survive, and liquid water flows on our special planet. These conditions do not exist anywhere else in our Solar System, but it is possible that there are livable planets circling other stars.

Q: WILL EARTH ALWAYS BE A GOOD HOME FOR LIVING THINGS?

Answer: As long as we take care of our planet, it will be a good home for living things. Humans are the most intelligent life-form on Earth, so we should be the caretakers of our home. We must keep the air and water clean. We must use resources such as fuel, water, and land wisely so that there will be plenty for all. Also, we must remember that the plants and animals that share the Earth with us are no less important than we are. Earth is home to us all.

Here are several other questions to consider about space.

WHAT IS THE DIFFERENCE BETWEEN A METEOROID AND A METEORITE?

WHAT CAUSES CRATERS TO FORM?

WHAT IS A BLACK HOLE?

WHAT CAUSES AN AURORA?

WHAT IS A GALAXY?

WHY ARE SOME STARS RED AND OTHERS BLUE?

DO SCIENTISTS KNOW HOW FAR AWAY THE STARS ARE?

DO OTHER STARS HAVE PLANETS, TOO?

WHAT IS A SUPERNOVA?

ARE THERE CLOUDS IN SPACE?

HOW LONG WOULD IT TAKE A SPACECRAFT TO GET TO THE MOON?

HOW DOES THE SPACE SHUTTLE REACH OUTER SPACE?

These books will help you discover the answers:

Harris, Alan, and Weissman, Paul: *The Great Voyager Adventure,* Englewood Cliffs, New Jersey, Julian Messner, 1990.

Herbst, Judith: *The Golden Book of Stars and Planets,* Racine, Wisconsin, Western Publishing Co., 1988.

Jobb, Jamie: *The Night Sky Book,* Covelo, California, Yolla Bolly Press, 1977.

Krupp, E. C.: *The Big Dipper and You,* New York City, Morrow Junior Books, 1989.